The Medal With a Heart

By Karen Kulinski

Illustrations by Karen Gruntman

M.T. Publishing Company, Inc.™

P.O. Box 6802
Evansville, Indiana 47719-6802
www.mtpublishing.com

Graphic Designer:
Amanda Reyher
M.T. Publishing Company, Inc. ™

Library of Congress Control Number: 2018955368

ISBN: 978-1-945306-89-1

Printed in the United States of America

Dedication

To my writer friends, Esther Hershenhorn and Peggy Archer, without whose help and expertise this book wouldn't have been published.

In truth, it takes a village to help a writer find success, and I am so grateful to my village:

Sharon Vargo and Angie Karcher who put me on the right path to finding a publisher who believed in my story, Mark Thompson at M.T. Publishing Company, Inc.

Peggy Reiff Miller who directed me to the absolutely perfect illustrator for this book.

Karen Gruntman whose wonderful illustrations brought my story to life.

And to my husband, Alan, who has walked beside me every step on my writer's journey.

A special note of thanks must go to Peter Bedrossian, at the National Purple Heart Hall of Honor, and Christopher Doerrer, at Washington's Headquarters State Historic Site, for all the help they gave me in making sure that this book is historically correct.

Use of the image of the Badge of Military Merit is courtesy of the New Windsor Cantonment, New Windsor, NY/ State Office of Parks, Recreation and Historic Preservation. The author is grateful for all their help.

Karen Kulinski

For L.R.F. my favorite Purple Heart recipient, my father.

K.S.G.

The Story of the Purple Heart Medal

Every medal tells a story.
Of bravery. Of hard work.

Of great achievement.

The story of The Purple Heart Medal begins almost 250 years ago with George Washington and the Revolutionary War.

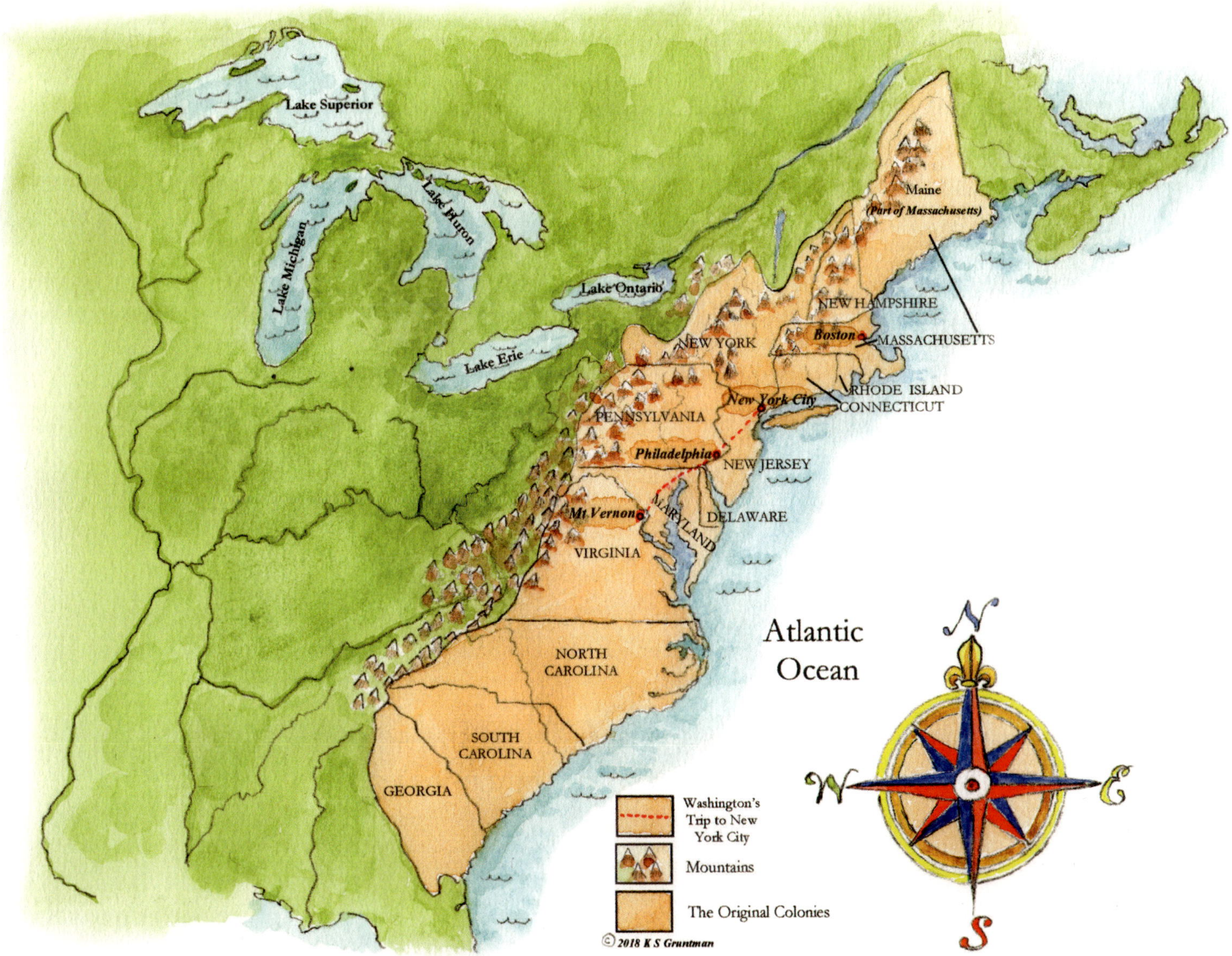

Back in 1775, America was made up of 13 different English colonies. But the Americans weren't getting along very well with England.

More and more, the colonies were being taxed -- on sugar, tea, even on paper documents like newspapers, wills and bills of sale.

Americans were English citizens, but they were not allowed to vote for representatives to England's government, which passed the tax laws.

"Taxation without representation is tyranny!" the Americans cried.

Feeling they were being treated unfairly, the colonies banded together to break free from England. They chose a man from Virginia, George Washington, to lead them.

Washington had bravely commanded Virginia's colonial troops during battles with the French army and local Indian tribes. Now he would lead soldiers from all the colonies to fight for America's freedom.

©KSG '18

The Revolutionary War was long, lasting eight years. Washington and his soldiers had to fight in many battles.

Often the men didn't have enough to eat.

A lot of them didn't have warm uniforms and sturdy boots to keep their feet from freezing.

Many nights the army had to sleep outside without blankets to cover them.

©2018
K.S.GRUNTMAN

General Washington wanted to do something special to thank his soldiers for all they had gone through to help free America.

He couldn't give his soldiers money. He couldn't give his soldiers land. What could he do for his men? Finally, Washington had an idea!

Up till now, only officers were given awards. So, the General decided to create a special award for enlisted men who had performed exceptionally brave or honorable actions during the war.

The award would be in the shape of a heart, to show his men how he felt about them.

It would be made from purple cloth, a color worn by kings.

Across the front of the heart the word MERIT was stitched, signifying the wearer had done something exceptional.

The award was called the Badge of Military Merit.

Washington announced the new Badge in his General Orders written on August 7, 1782, and all his officers received a copy. They were instructed to give Washington the names of any soldiers whom they felt had performed an exceptional action during the war.

"The road to glory in a patriot army and a free country is thus open to all!" Washington declared.

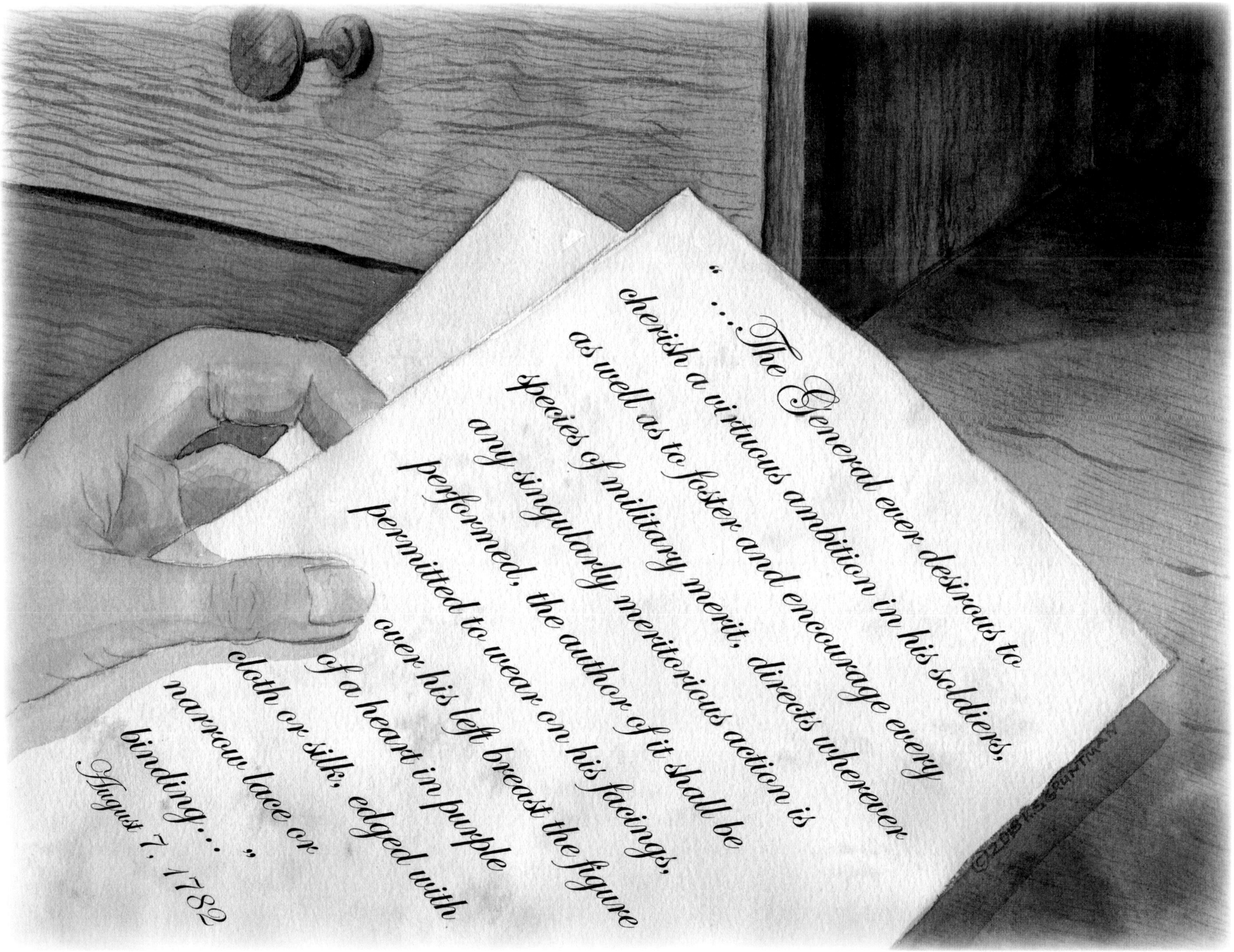
"...The General ever desirous to
cherish a virtuous ambition in his soldiers,
as well as to foster and encourage every
species of military merit, directs whenever
any singularly meritorious action is
performed, the author of it shall be
permitted to wear on his facings,
over his left breast the figure
of a heart in purple
cloth or silk, edged with
narrow lace or
binding..."
August 7. 1782

The General dreamed of giving his special award to many of his soldiers, but months and months passed without him hearing from any officer. Then word came in early April,1783, that England's King George III had decided not to fight anymore.

Frustrated, Washington called a committee of officers together and, finally, he received two names. Later, the General himself added a third name to the list.

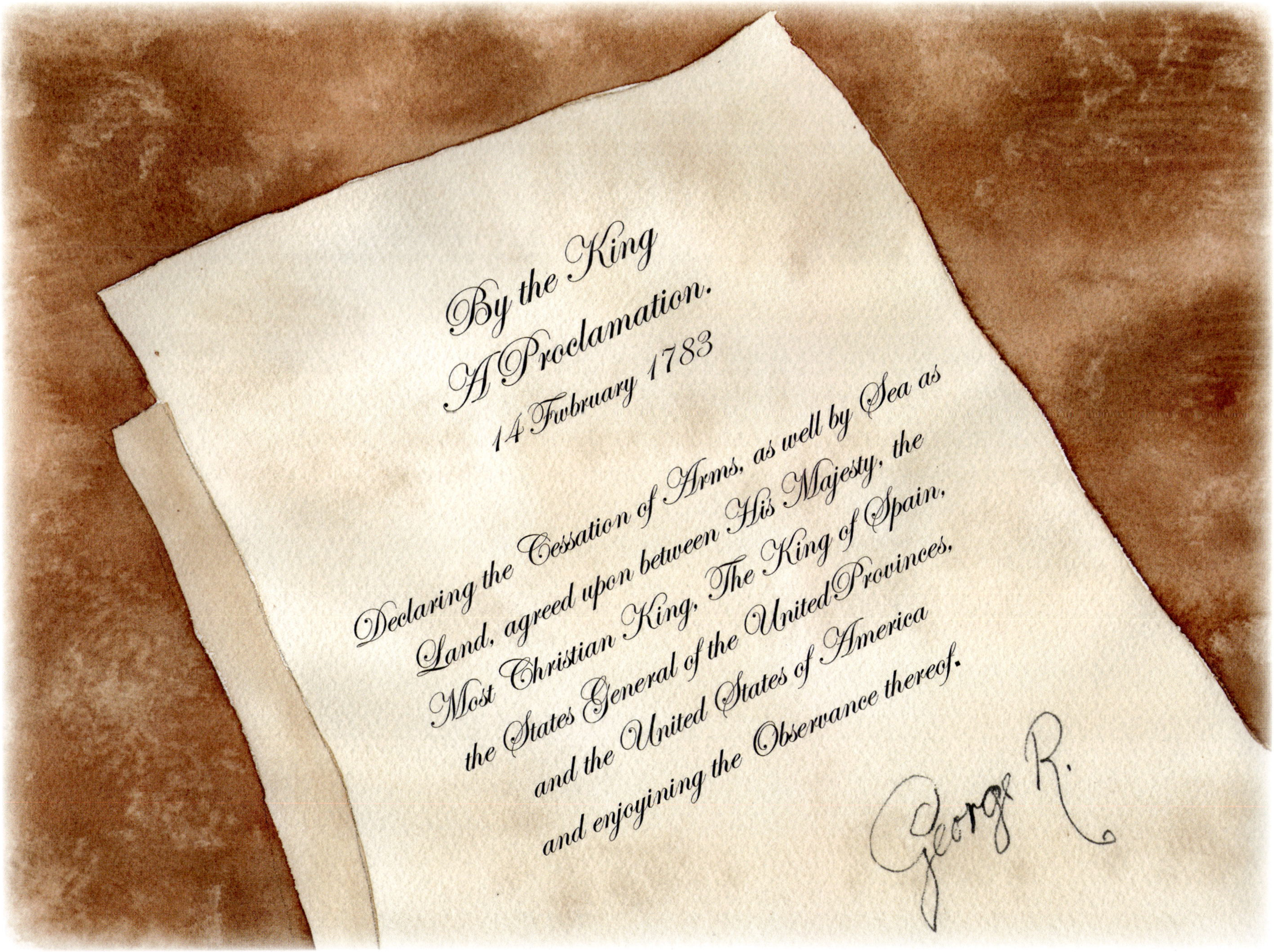
By the King
A Proclamation.
14 Fwbruary 1783
Declaring the Cessation of Arms, as well by Sea as
Land, agreed upon between His Majesty, the
Most Christian King, The King of Spain,
the States General of the United Provinces,
and the United States of America
and enjoyining the Observance thereof.
George R.

On May 3, 1783, Sergeants Elijah Churchill and William Brown, both from Connecticut regiments, received their Badges of Military Merit at Washington's Headquarters. Another Connecticut sergeant, Daniel Bissell, received his award there on June 10th

With the fighting now ended, the soldiers just wanted to go home. So, the Badge of Military Merit faded into history.

©2018
K.S. GRUNTMAN

George Washington finally made it home to his beloved Mount Vernon just in time for Christmas. He was America's hero, but he now looked forward to simply being a farmer again.

Washington didn't stay a farmer for long.

Six years later, the people of the new United States of America elected George Washington their first president.

On April 16, 1789, he wrote in his diary, “About ten o’clock I bade adieu to Mount Vernon, to private life ... to render service to my country in obedience to its call ....”

©2018
K.S. GRUNTMAN

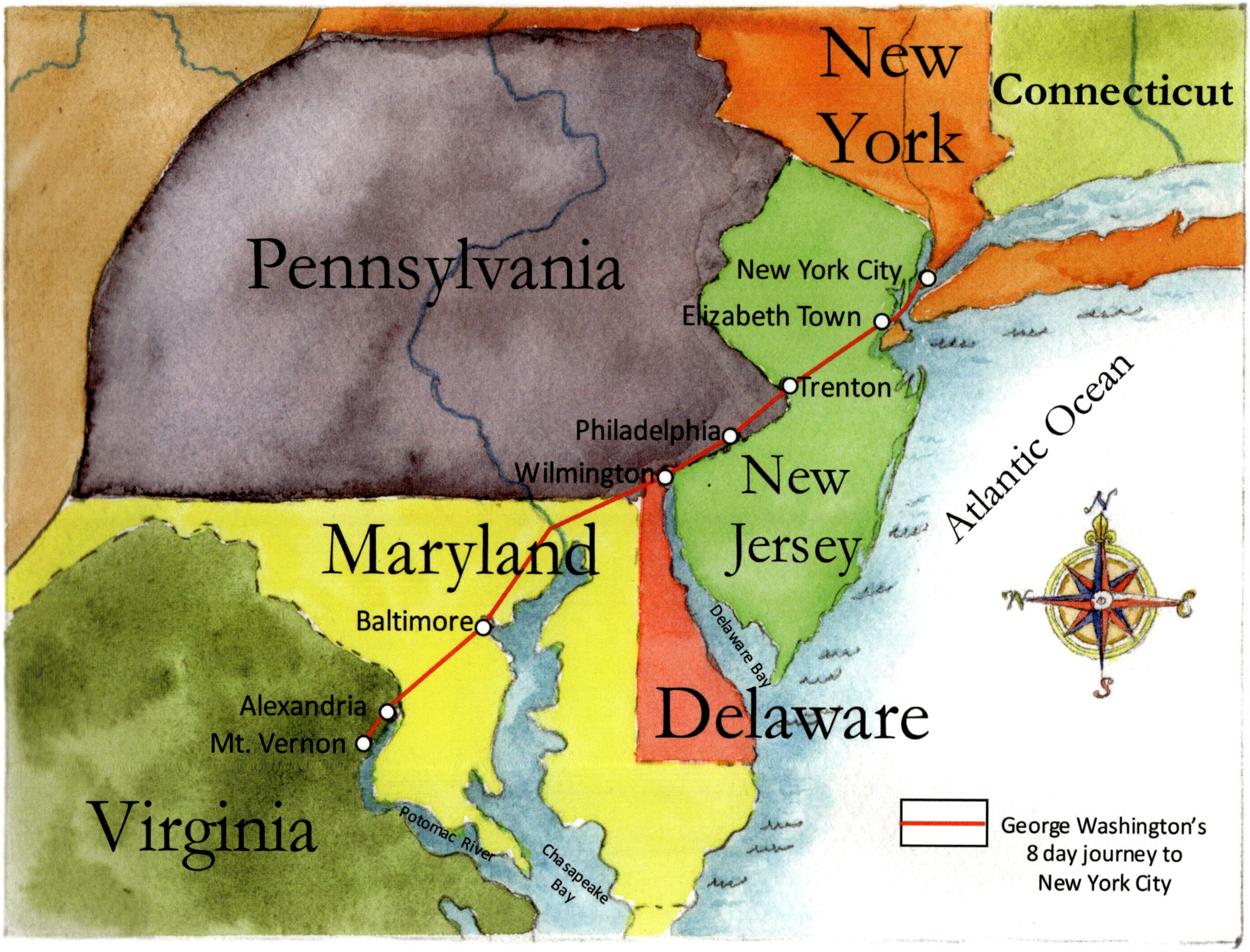

Then he rode off to America's new capital, New York City, to help build a nation.

Federal Hall, New York City

Almost 150 years later, in 1932, America wanted to honor George Washington on his 200th birthday. So, they gave him the best birthday present ever.

They brought his Badge of Military Merit back to life ... with a few changes.

It was still heart-shaped, but the new award was not made from cloth. It was made of brass, coated with gold.

The front of the medal was purple, with an image of George Washington on it.

On the back were the words, “For Military Merit.”

They gave it a new name, too -- The Purple Heart.

©2018 K.S.G.

FOR
MILITARY
MERIT
Lowell R.Falk

The new Purple Heart Medal was to be awarded to any member of the United States Armed Forces — the Army, the Navy, the Marines, the Air Force, the Coast Guard — who was wounded or killed fighting for our country.

©2018 K.S.G.

George Washington dreamed of awarding his Badge of Military Merit to many American soldiers.

Estimates are that since 1932, the Purple Heart has been awarded to approximately 1,800,000 members of the Armed Forces.

Washington's dream finally came true.

S.ARMY
©2018 K.S. Gruntman

Author's Notes

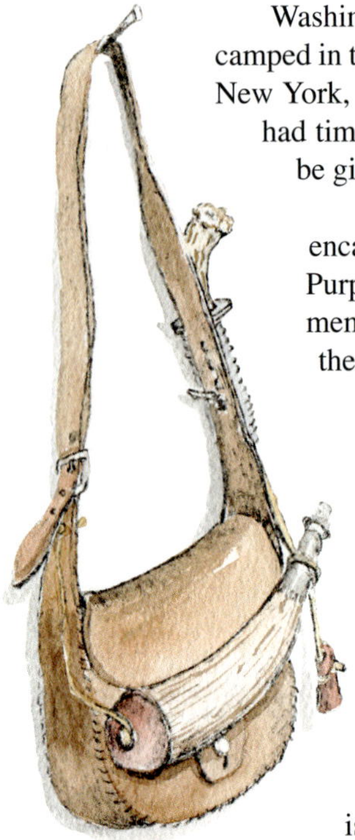

Washington's army spent the last year of the Revolutionary War camped in the wilderness keeping the British army cooped up in the city of New York, 60 miles to the south. With no fighting going on, Washington had time to create his Badge of Military Merit, the first award ever to be given to enlisted soldiers.

Today, in New Windsor, New York, on the site of that final encampment of the northern Continental Army, stands the National Purple Heart Hall of Honor. Opened in 2006, it memorializes those members of the American Armed Forces who have been awarded the Purple Heart.

Sergeant Churchill's Badge of Military Merit -- the only one of the three original badges known to survive -- is proudly displayed there.

About the Author

When Karen Kulinski first learned about the origins of the Purple Heart Medal, she immediately thought, "This is a story that's got to be told." It took some years and a lot of effort, but she finally made it happen. When Karen's not writing, she can be found at the railroad museum she helped create in her hometown of Griffith, Indiana, or spending time with her husband, Alan, and their grandchildren. For more information on Karen, her books and her museum, check out her website: www.karenkulinski.com

About the Illustrator

Karen Gruntman studied art at Central Michigan University and children's book illustration with Caldecott Award-winning author/illustrator Uri Schulevitz. Illustrating a number of other books has been a real joy for her, among them was *What Is That Thing? Who's Stuff Is This?* by John Gile. She lives with her husband, Dennis, and little dog, Bristol, in Goshen, Indiana. Spending time with her family, world travel, knitting, reading and flower gardening are among her favorite activities.

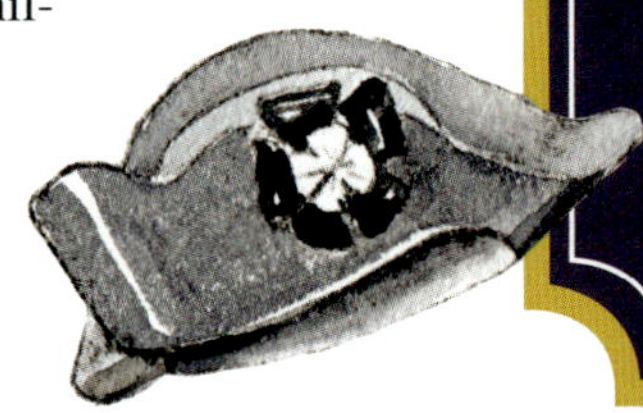

Purple Heart Mysteries

Only the names of the three Connecticut Sergeants in this story are listed in Washington's General Orders as having received the Badge of Military Merit during the Revolutionary War.

However, the Military Order of the Purple Heart has uncovered Revolutionary War records indicating that three other Continental Army soldiers received the award, not for exceptional acts but for years of honorable service.

So we are left to wonder if, in the frantic last days of the Revolutionary War, were others awarded Washington's Badge of Military Merit? The names of all Badge recipients and their heroic deeds were to be put down in a "Book of Merit," but this book has never been found.

In the 1920's, a small faded lavender silk heart stitched to a fragment of an old wool uniform was found in a New Hampshire barn. The heart does not resemble Sergeant Churchill's authenticated Badge, but George Washington's General Orders state that the badge was to be made from "purple cloth or silk."

This lavender silk heart is now on display at the American Independence Museum in Exeter, New Hampshire.

Sergeant Elijah Churchill's Story

On a dark November night in 1780, Sergeant Elijah Churchill, 2nd Continental Dragoons, was at sea. He was part of a raiding party headed for Fort St. George on Long Island to destroy British supplies.

After landing, the Americans marched 20 miles in the dark and reached the fort just before dawn.

The raiding party was split into three groups, and Churchill was put in charge of 16 men. He led them through sharpened tree branches and over piled-up trees. Then the British started firing.

Undaunted, Churchill and his men scrambled through a wide ditch and overwhelmed the enemy. Soon the fort was in American hands.

After burning the fort, the supplies, and a ship docked near the shore, the men marched 20 miles back to their boats.

Eleven months later, Sergeant Churchill was wounded while leading another raid that destroyed British weapons and took 21 prisoners.

After the War, Churchill and his family moved to Massachusetts. His Badge of Military Merit was handed down to a grandson and is now on display at the National Purple Heart Hall of Honor in New Windsor, New York.

Sergeant William Brown's Story

On a cool October night in 1781, Sergeant William Brown, 5th Continental Dragoons, faced almost certain death. He was about to lead a surprise raid against the British at Yorktown, Virginia.

So the surprise wasn't foiled, he and his men carried unloaded weapons.

A thick pile of trees, their branches sharpened and sticking out in all directions, surrounded the town. Behind the trees, were entrenched British marksmen.

Armed only with bayonets, Brown led his men through the sharpened branches and over the trees. The murderous hail of musket fire didn't stop them from attacking the soldiers in the trenches.

After ten minutes of intense fighting Brown and his men were inside the town. The rest of the raiding party followed and soon captured the surprised British soldiers.

After the war, Brown homesteaded a farm on the Ohio frontier. His Badge of Military Merit was handed down through his family. In 1924, it was stolen and has never been found.

Sergeant Daniel Bissell's Story

In the summer of 1781, Sergeant Daniel Bissell, 2nd Connecticut Continental Line Infantry, posed as a deserter and joined the British Army in New York city. He was there under direct orders of George Washington to spy on the British.

Bissell served under Benedict Arnold. His duties took him all over the city, allowing him to gather information on British troop numbers and their movements.

Every moment of every day, Bissell's life hung by a thread. One mistake, one wrong decision, and he could have been discovered to be a spy and executed.

After almost a year in New York, Bissell escaped and made his way through the forest to an American patrol. Then he faced a new problem.

The Americans thought he was a British spy and threatened to execute him!

Finally, Bissell convinced his captors to take him to Washington, where he could finally tell the General everything he had learned about the British.

Daniel Bissell settled with his family in northern New York state. In 1813, his Badge of Military Merit was destroyed in a fire.

Camping Out, Continental Army Style

(Drawing by Tim Kulinski)

During the Revolutionary War, the British and American armies in the north took the winter off. Cold, snowy weather made it hard to move troops and sometimes impossible to get needed supplies to the men and horses. So the armies camped out during the winter.

Soldiers lived in log huts they built themselves, like the one pictured here.

At the Continental Army's final encampment in New Windsor, New York, the winter of 1782-83, 600 huts were built to house the more than 7,000 men, plus 500 women and children who were accompanying the army.

The women, many war widows with nowhere to go, served as cooks, nurses, seamstresses and washerwomen for the troops.

General Washington and other high-ranking officers did not camp out. Instead, they were headquartered in private homes in nearby towns, and wives were invited to stay with their husbands over the winter.

Martha Washington came most winters to be with George. The last one of the War, Washington was headquartered in an 8-room house a few miles away from the New Windsor camp. It was here, at Hasbrouck House, in Newburgh, New York, shown to the right, that the Badge of Military Merit was created.

This house, and the grounds of the final encampment — now called the New Windsor Cantonment — are New York State historical sites. Sharing the Cantonment grounds is the Purple Heart Hall of Honor, memorializing awardees and displaying the only authenticated surviving Badge of Military Merit.

(Drawing by Tim Kulinski)

Firecakes

Washington knew a well-fed army made for the best fighters. So he ordered each soldier be given portions of beef, pork or salt fish; bread or flour to make bread; peas, beans or other vegetables; milk; rice; Indian meal; and cider.

As the war progressed, soldiers were lucky to receive half the food they were promised. Things often got so bad that soldiers ended up having only flour and water. When that happened, they made firecakes.

The men mixed the flour and water together, adding salt if they had it, and formed the mixture into the shape of a flattened cake. This cake was then baked on a rock set in the fire, and sometimes actually put right in the fire until blackened.

On a good day, the cakes were tasteless; on bad days, the flour would have contained weevils or maggots which may or may not have been removed before the cakes were made. A firecake was very hard, not very tasty, but eating the blackened biscuit was better than going to sleep hungry.

Timeline

February 22, 1932

The War Department announces: *By order of the President of the United States, the Purple Heart established by General George Washington at Newburgh, August 7, 1782, during the War of the Revolution, is hereby revived out of respect to his memory and military achievements.*

By order of the Secretary War:
Douglas MacArthur
General, Chief of Staff

On May 28, 1932

137 World War I veterans were conferred their Purple Hearts in a farm field on Temple Hill, in New Windsor, New York. Temple Hill was the site of the New Windsor Cantonment, which was the final encampment of the Continental Army in the winter of 1782-1783.

December 3, 1942

By Executive Order President Franklin Roosevelt authorizes the Purple Heart for all branches of service and authorizes posthumous awards back-dated to December 6, 1941.

November 12, 1952

Executive Order 10409 authorizes posthumous Purple Heart awards to Navy, Coast Guard or Marine Corps personnel killed on or After April 5, 1917. (The Army and the Air Force are not formally included until 1962.)

April 25, 1962

Executive Order 11016 authorizes civilian Purple Hearts for those under competent military authority; it also authorized posthumous awards to those killed on or after April 5, 1917, upon application by their next of kin.

February 23, 1984

Executive Order 12464 authorizes the Purple Heart to be awarded for acts of terror, as well as for wounds or death resulting from U.S. Armed Forces Personnel acting as part of a peacekeeping force outside of the United States or its territories.

November 30, 1993

Purple Hearts may be awarded for wounds or death resulting from "friendly fire" (unless it is from willful misconduct.) U.S. Code 10, section 1129, per PL 103-160

February 10, 1996

PL104-106, Section 521 expands Purple Heart eligibility to POWs wounded during capture or during captivity prior to April 25, 1962. (Policy interpretations had considered, and awarded the Purple Heart, on a case by case basis to POWs captured after April 25, 1962.)

May 19, 1998

Effective this date, The Purple Heart is limited to American military personnel, and civilian awards are eliminated.

October 1, 2008

The Department of Defense authorizes the Purple Heart for POWs (after December 7, 1941) who subsequently die in captivity. Information is from Memo this date to secretaries of military departments.

April 28, 2011

The Department of Defense announces a standard which allows for a more standard way to evaluate a wounded individual for a Purple Heart resulting from a "non-penetrating wound."

February 6, 2015

The Department of Defense announces that eligibility has been extended to those wounded or killed by certain kinds of domestic terrorist activities.

Purple Heart - Did You Know?

The Purple Heart is the oldest existing United States military award, with the exception of the "Fidelity Medallion." This medallion was a one-time award given to three soldiers for a specific event, that being the capture of Major John Andre, the British spy who helped facilitate Benedict Arnold's treason.

The Purple Heart differs from all other military decorations in that an individual is not "recommended" for the decoration; rather he or she is entitled to it upon meeting specific criteria. Any member of the Armed Forces who sheds blood in defense of the nation is automatically awarded the Purple Heart. If a member of the U.S. military believes that he or she is eligible for the Purple Heart, but through unusual circumstances no award was made, may submit an application through military channels.

In military terms, the award had "broken service," as it was ignored for nearly 150 years until it was re-introduced on February 22, 1932, on the 200th anniversary of George Washington's birth.

Originally the Purple Heart was awarded for "meritorious service" like the Badge of Military Merit. Being wounded was considered to be one type of merit.

With the creation of the Legion of Merit in 1942, the requirements for the Purple Heart were changed. From then on, the Purple Heart would be awarded to anyone who was wounded or killed during military service in the U.S. Armed Forces. But further changes occurred over the years, which are listed in the Purple Heart timeline on page 37.

No record survives today of the identity of the first individual to be awarded the revived and redesigned Purple Heart.

A Purple Heart is awarded for the first wound suffered, but for each subsequent wound is award an oak leaf cluster or 5/16 inch star to be worn on the ribbon in lieu of another medal.

The Purple Heart isn't just symbolic. Being awarded a Purple Heart also means access to some important military and civilian benefits, among them Forever GI Bill Purple Heart benefits, College Tuition waivers in some states, Purple Heart Scholarship programs through the Military Order of the Purple Heart, and Purple Heart license plates.

President John Fitzgerald Kennedy was the only United States president awarded the Purple Heart.

The Purple Heart Medal contains the shield of the Washington coat of arms at the top of the pendant.

In his own day, Washington had been appointed and served as "General and Commander in chief of the Continental Army and wore a three-star insignia and his rank was informally called a "three star general" when he retired. However, with the passage of time, Washington was outranked by four- and five-star generals of the Civil War, World Wars I and II. To rectify this, on July 4, 1976, as part of the Bicentennial Events, Washington was posthumously promoted to the full grade of General of the Armies of the United States by Congress, which further declared that no officer of the U.S. Armed Forces should ever outrank George Washington.

Beatrice Mary MacDonald

Beatrice Mary MacDonald is the first known woman to be awarded the Purple Heart. Beatrice, an Army nurse, was injured in World War I while working in a British Clearing Hospital four miles behind the lines in Belgium when a German aircraft bombed the hospital. The injuries caused her to lose her right eye, but Beatrice remained in the Army and served throughout the war.

Lieutenant Annie G. Fox

The first woman to be awarded the Purple Heart for singularly meritorious action was Lieutenant Annie G. Fox, chief nurse in the Army Nurse Corps at Hickham Field during the Japanese attack on Pearl Harbor on December 7, 1941. At this time it was not necessary to be wounded to be awarded the Purple Heart. Fox was awarded the medal for her outstanding performance as head nurse during the bombardment of Pearl Harbor, working under extreme conditions, and embodying calm and courage.

Rita G. Palmer

The first woman awarded the Purple Heart who was wounded in action was Rita G. Palmer. Rita was a young nurse from Hampton, New Hampshire, who joined the Army Nurse Corps in 1940. She served in the field hospitals in the Philippines until the Americans surrendered in 1942 when she was wounded and taken prisoner. She was a prisoner in Santa Tomas for three years until the American liberation in 1945.

Purple Heart Stamp

(Drawing by Tim Kulinski)

The Purple Heart stamp was first issued in 2003 and became a Forever stamp in 2011. Before 2011, every time the price of postage increased, letter campaigns were undertaken to keep the Purple Heart stamp. The Purple Heart stamp joined the Liberty Bell stamp, which had been the only Forever stamp up to that point, as symbols of freedom, sacrifice, and remembrance.

Elizabeth Will

Elizabeth Will was a heraldic specialist for the U.S. Army and was put in charge of the redesign of the modern Purple Heart award. Associated with the Army's heraldic program for 39 years, she designed the Distinguished Flying Cross, the official flag for the Vice President and many seals, badges, insignia and flags for the Department of Defense and other governmental agencies. At her retirement, she was presented with the Meritorious Civilian Service Award, one of the Army's highest awards for civilian employees.

The National Purple Heart Hall of Honor

The National Purple Heart Hall of Honor in New York's Hudson River Valley offers visitors an incredible journey through military history as well as reminders of human sacrifice and the cost of freedom.

Each branch of service is represented, and the range of conflicts for which the Purple Heart has been available are depicted. Four cases reflect the site's connection from the American Revolution to Temple Hill Day, May 28, 1932, when some of the first Purple Heart Medals were awarded.

The theme "Every Purple Heart Has a Story to Tell" is represented by personal story cases and other displays. Visitors can search two interactive kiosks by name, branch of service, conflict, or theme to hear Purple Heart recipients relate their experiences. The Hall of Honor greatly appreciates the assistance of visitors in building their Roll of Honor database of Purple Heart recipients.

For more information, check out the National Purple Heart Hall of Honor website.

The New Windsor Cantonment

In October 1782, General Washington moved his northern army to New Windsor to set up winter quarters. It was at the Cantonment that cease-fire orders were issued by Washington on April 19, 1783, ending the War of Independence.

From mid-April through October visitors will find costumed staff demonstrating musket drills, blacksmithing, military medicine and other 18th century skills, with additional demonstrations in July and August. Buildings include the reconstructed "Temple of Virtue," and perhaps the only surviving example of timber construction by the Continental Army.

Special events include reenactors each April; Memorial Day and Independence Day activities; and a Children's Day to honor the children who lived at the Cantonment.

For more information, check out the New Windsor Cantonment website.

The Military Order of the Purple Heart

The Military Order of the Purple Heart (MOPH), begun in 1932 and chartered by Congress in 1958, is composed of military men and women who were awarded the Purple Heart Medal for wounds suffered in combat.

Although membership is restricted to the combat-wounded, MOPH provides support to all veterans, has scholarship programs for members and their families, maintains continuous liaison with Congress on key issues related to veterans, and Chapters throughout the country host a myriad of local activities.

For more information on the Military Order of the Purple Heart, contact their website.

Washington's Headquarters State Historic Site

George Washington moved into Hasbrouck House, his winter headquarters in Newburgh, New York, on March 31st, occupying the 8-room private home until August,1783. Hasbrouck House later became the nation's first publicly owned historic site and visitors can tour the rooms where American history was made.

It was at Hasbrouck House that George Washington created the Badge of Military Merit, the forerunner of the Purple Heart.

Check their website for hours of operation, tours, and special events throughout the year.

"The Story of the Purple Heart: The Medal of Honor of the Revolution." *Daughters of the American Revolution Magazine*, Vol. LVI, No. 2, 22 February 1922, pp. 63-70.

Wensyal, James W. "The Badge of Military Merit." *American History* Magazine, May 1982, pp. 26-31.

Raymond, Ray MBE, FRSA. "The Badge of Military Merit." Military Order of the Purple Heart, 2007.

Guide to the National Purple Heart Hall of Honor at the New Windsor Cantonment State Historic Site, New Windsor New York. New Windsor, New York.

Bedrossian, Peter. Personal interview.

"Updated History of the Purple Heart." http://www.mophhq.org.

Doerrer, Christopher. Citations from the facility's primary documents in regard to the Badge of Military Merit, Washington's Headquarters State Historic Site, Newburgh, New York.

Writings of Washington, Vol. 24, pp. 487-488; Vol. 26, pp. 363-364, pp. 481-482.